LEARN ABOUT

INSECTS

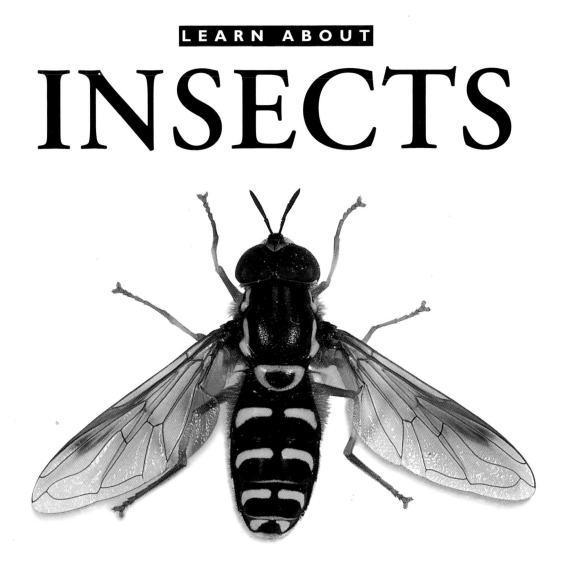

JEN GREEN

LORENZ BOOKS

This edition published in the UK in 1998 by Lorenz Books

© Anness Publishing Limited 1998

Lorenz Books is an imprint of
Anness Publishing Limited, Hermes House
88–89 Blackfriars Road, London SE1 8HA

This edition distributed in Canada by Raincoast Books,
8680 Cambie Street, Vancouver, British Columbia V6P 6M9

ISBN 1 85967 644 8

A CIP catalogue record for this book is available
from the British Library

Publisher: Joanna Lorenz
Managing Editor, Children's Books: Sue Grabham
Project Editor: Sophie Warne
Text Editor: Charlotte Evans
Consultant: Michael Chinery
Children's Photographer: John Freeman
Nature Photographer: Robert Pickett
Stylist: Melanie Williams
Designer: Caroline Grimshaw
Illustrator: Alan Male

Printed and bound in China

10 9 8 7 6 5 4 3 2 1

The Publishers would like to thank the following children, and their
parents, for modelling in this book – Rodney Ammah, Dolly
Batkhismig, Joshua Cooper, Charlene Da-Cova, Rachel Greiner,
Ion Kojokari, Armani McKenzie, Alejandro Otalvora,
Sussy Quirke, Mark Stafford.

Thanks also go to Sarah Wood.

INSECTS

CONTENTS

INCREDIBLE INSECTS

Swallowtail butterfly

INSECTS are the most successful animals on Earth. More than a million species are known, with many more still to be discovered. Insects make up three-quarters of all animal species in the world. They are found in steamy rainforests, on top of mountains and in the middle of baking deserts. A few kinds are harmful, spreading disease and damaging crops or buildings. Many more, however, are helpful – they pollinate plants and help fertilize the soil. As many as 10,000 insects can live on a single square metre of the Earth's surface, making them an important food source for many animals. Insects are invertebrates, which means they do not have a backbone – unlike birds, reptiles and mammals. They have three pairs of legs and most adult insects have one or two pairs of wings. To help identify insects, experts divide them up into orders. All the species in one order have similar life cycles and share certain features, such as the shape of their wings or mouthparts. The largest orders are shown here. Each one contains many thousands of species.

Butterflies and moths (*Lepidoptera*)
The swallowtail butterfly belongs to the *Lepidoptera* order, which has about 150,000 species. The wings of butterflies and moths are covered with tiny scales that overlap like roof tiles. The order name means 'scale wings'.

Bugs, hoppers and aphids (*Hemiptera*)
This shieldbug is one of about 80,000 species of bugs, hoppers and aphids that all belong to the *Hemiptera* order. Their order name means 'half wing'. This refers to the front wings of many larger bugs, which are tough at the base but soft at the tip.

Shieldbug

Greenbottle

Flies (*Diptera*)
There are about 100,000 species of *Diptera*, including this greenbottle. *Diptera* means 'two wings'. Unlike most insects, flies have only one pair of wings. Instead of rear wings, flies have tiny balancing organs called halteres.

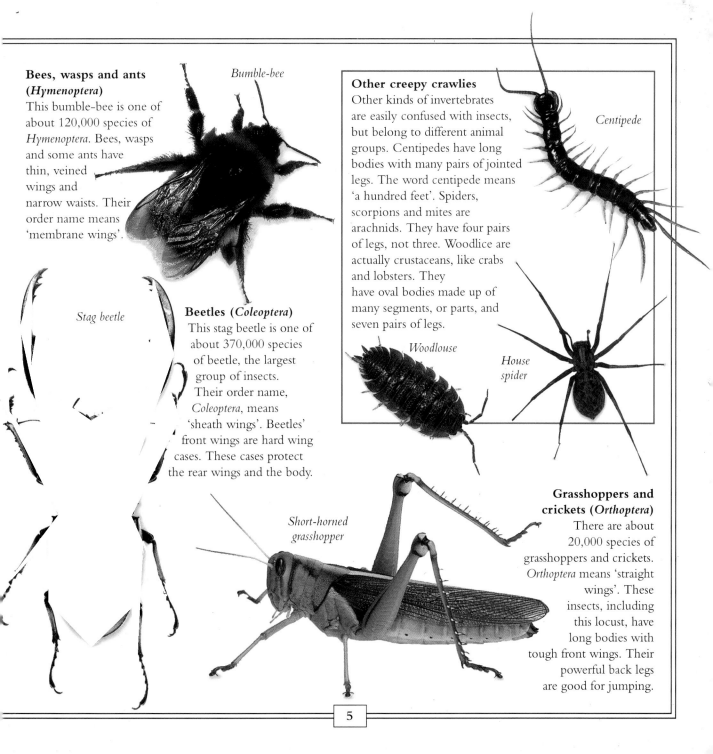

Bees, wasps and ants (*Hymenoptera*)

This bumble-bee is one of about 120,000 species of *Hymenoptera*. Bees, wasps and some ants have thin, veined wings and narrow waists. Their order name means 'membrane wings'.

Bumble-bee

Stag beetle

Beetles (*Coleoptera*)

This stag beetle is one of about 370,000 species of beetle, the largest group of insects. Their order name, *Coleoptera*, means 'sheath wings'. Beetles' front wings are hard wing cases. These cases protect the rear wings and the body.

Other creepy crawlies

Other kinds of invertebrates are easily confused with insects, but belong to different animal groups. Centipedes have long bodies with many pairs of jointed legs. The word centipede means 'a hundred feet'. Spiders, scorpions and mites are arachnids. They have four pairs of legs, not three. Woodlice are actually crustaceans, like crabs and lobsters. They have oval bodies made up of many segments, or parts, and seven pairs of legs.

Centipede

Woodlouse

House spider

Grasshoppers and crickets (*Orthoptera*)

There are about 20,000 species of grasshoppers and crickets. *Orthoptera* means 'straight wings'. These insects, including this locust, have long bodies with tough front wings. Their powerful back legs are good for jumping.

Short-horned grasshopper

LOOKING FOR INSECTS

Never go insect watching on your own – always take an adult with you. On longer trips you may need food, drink and warm waterproof clothing.

INSECTS are everywhere, so they are easy to study. The best place to start is your local area, in a park or garden. You will find an amazing variety. Up to 300 species of beetles can be found in an average city garden, with as many kinds of moths and several butterflies. The same small area may contain up to 200 kinds of flies, 90 different bugs and many species of bees, ants and wasps. So there is plenty to see! Different types of insect live and feed in long grasses, flower beds and hedges, under logs and stones and in the soil. Find out which species prefer these different habitats, or homes, by carrying out a survey. Peg out study areas in four different sites and see how many species you can find in each. When looking for insects, remember to replace any stones and logs you turn over. Try not to damage plants by breaking stems and flowers.

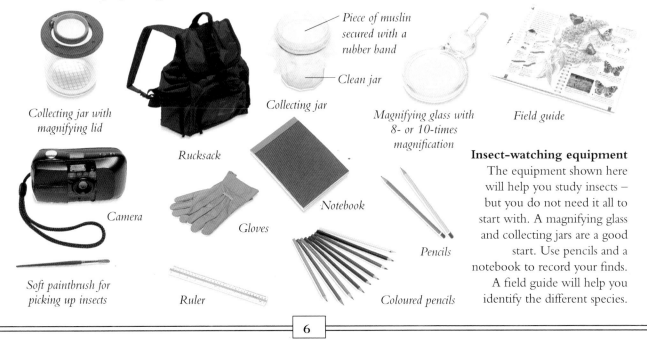

Collecting jar with magnifying lid

Rucksack

Camera

Piece of muslin secured with a rubber band

Clean jar

Collecting jar

Gloves

Notebook

Magnifying glass with 8- or 10-times magnification

Field guide

Pencils

Soft paintbrush for picking up insects

Ruler

Coloured pencils

Insect-watching equipment
The equipment shown here will help you study insects – but you do not need it all to start with. A magnifying glass and collecting jars are a good start. Use pencils and a notebook to record your finds. A field guide will help you identify the different species.

Insect survey

1 Find an area of long grass. Wearing gloves, use the tent pegs and string to mark out a square measuring one metre on each side.

MATERIALS

You will need: gardening gloves, four tent pegs, string, tape measure, collecting jar, magnifying glass, pen or pencil, notebook, field guide, coloured pencils.

2 What insects can you find inside the square? Use a collecting jar and magnifying glass to study them. Describe your finds in a notebook.

3 Now mark out a square metre in a different place. Try an area with flowers or a hedge. You may find aphids and ladybirds on plant stems, and shieldbugs under leaves.

6 Use your field guide to identify your finds. How many species did you find in each area? Make a chart to record your survey's results.

4 Move a fallen log to see what kinds of insects live underneath. Make sure you wear gloves to protect your hands. You may find beetles under logs and earwigs under bark.

5 Still wearing gloves, carefully look under some stones. What kinds of insect prefer this habitat? You may find ground beetles or an ants' nest.

WHAT IS AN INSECT?

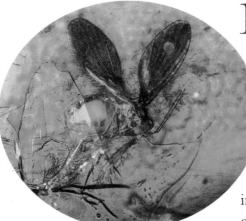

Bᴵᴿᴰˢ, reptiles and mammals all have internal skeletons to provide a framework for their bodies. Insects are different – they have their skeletons on the outside. Their soft body parts are protected by a hard case called an exoskeleton. This case is made of a tough, light material called chitin, which forms a waterproof barrier around the insect. Chitin prevents the insect from drying out and also prevents air from passing through. To let air inside an insect's body there are tiny holes, called spiracles, along each side. Unlike birds or mammals, insects are cold-blooded animals. This means the temperature of an insect's body is about the same as its surroundings. To warm up, an insect basks in the sunshine. When it gets too hot, it moves into the shade. In very cold weather insects find it difficult to move at all. Many adult insects die in winter. Others enter a deep sleep called hibernation and wake up again when it is spring.

Inside this piece of amber is a fossil fly. Amber is the fossilized sap from pine trees that grew millions of years ago. The fly was trapped in the sticky sap and preserved as the sap hardened and slowly turned to amber.

Insect fossil
Insects are a very ancient group of animals, as this 130-million-year-old dragonfly fossil shows. Insects have lived on Earth for over 400 million years, and were the first animals to fly 350 million years ago. Many species we know today developed about 100 million years ago, when flowering plants first appeared.

Modern insects
This modern dragonfly looks very similar to its prehistoric ancestor. Most of the earliest insect species, however, are now extinct.

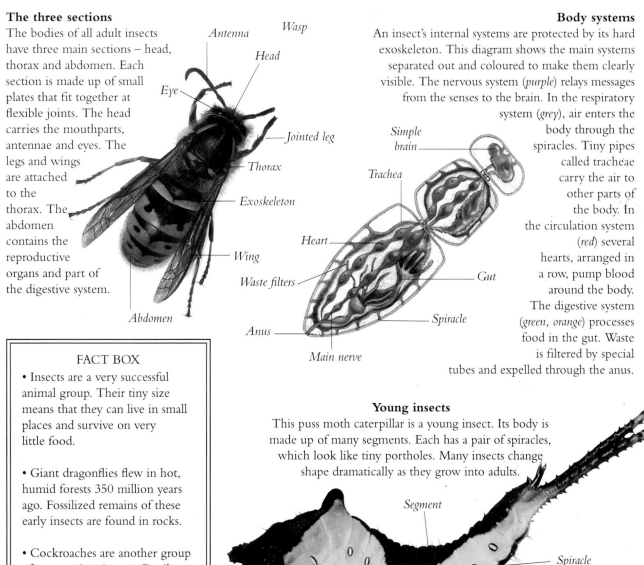

The three sections

The bodies of all adult insects have three main sections – head, thorax and abdomen. Each section is made up of small plates that fit together at flexible joints. The head carries the mouthparts, antennae and eyes. The legs and wings are attached to the thorax. The abdomen contains the reproductive organs and part of the digestive system.

Wasp

Antenna

Head

Eye

Jointed leg

Thorax

Exoskeleton

Wing

Abdomen

Body systems

An insect's internal systems are protected by its hard exoskeleton. This diagram shows the main systems separated out and coloured to make them clearly visible. The nervous system (*purple*) relays messages from the senses to the brain. In the respiratory system (*grey*), air enters the body through the spiracles. Tiny pipes called tracheae carry the air to other parts of the body. In the circulation system (*red*) several hearts, arranged in a row, pump blood around the body. The digestive system (*green, orange*) processes food in the gut. Waste is filtered by special tubes and expelled through the anus.

Simple brain

Trachea

Heart

Waste filters

Anus

Main nerve

Gut

Spiracle

Young insects

This puss moth caterpillar is a young insect. Its body is made up of many segments. Each has a pair of spiracles, which look like tiny portholes. Many insects change shape dramatically as they grow into adults.

Segment

Spiracle

Puss moth caterpillar

RECORDING INSECTS

You will need: magnifying glass, pencil, notebook, coloured pencils.

YOUR notebook is a vital piece of equipment. Use it to make notes of all the insects that you see. If you have a spiral-bound notebook, you could tear out pages and keep all the notes you make about an insect together. Write down the date, time, weather conditions and the place where you found the insect. Try drawing a map to show its location. Make rough notes and write them up neatly later. When you see an insect that you want to identify, try to find out which family it belongs to. A field guide will help with this. Ask yourself questions. What shape is the insect's body? Is it short and rounded, or long and slender? Does it have wings? If so, how many and what are they like? Does the insect have hard wing cases, or long legs? Now look at the mouthparts and antennae. The steps below will help you to make drawings of the insects you see. Insects are fragile and can be difficult to pick up and examine without harming them. A simple pooter will help you collect insects for study.

Drawing insects

1 Use the magnifying glass to study your insect closely. Start by drawing three ovals to show the insect's head, thorax and abdomen.

2 Can you see the insect's legs? Now copy the size and shape of its antennae. Draw in the eyes and add the outline of the wings.

3 Now draw any markings you notice on the insect's wings and body. Finish off by colouring in the drawing as accurately as you can.

Make a pooter

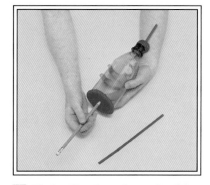

1 Cut off the bottom of the plastic bottle. Roll out one large and one small ball of modelling clay. Flatten out the large ball and mould it over the bottom of the bottle.

2 To make a filter, cut a short piece of straw. Secure a piece of muslin around the straw with a rubber band. Push the other end through the small lump of modelling clay.

3 Fit the filter into the neck of the bottle by moulding the clay. Make a hole in the bottom flap of clay with a sharp pencil. Fit a long straw into the hole you have made.

M A T E R I A L S

You will need: small plastic bottle, scissors, modelling clay, wide bendy straws, small piece of muslin, rubber band, sharp pencil.

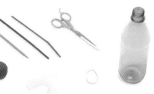

4 Go outside and look for a small insect you want to study. Do not try to suck up big insects as you will damage them! Aim the end of the long straw over the insect. Suck in on the short straw to draw the insect safely into the pooter.

5 When you have finished studying your insect, take it back to where you found it. Release the insect by carefully removing the bottom flap of modelling clay and shaking the insect out.

SEEING AND HEARING

FOR an insect the world is a dangerous place. There are many enemies eager to kill and eat it. In order to survive, an insect must find food and escape its enemies. It is helped by the senses of sight, hearing, touch, taste and smell. But insect senses are not the same as ours. Most species have three simple eyes, called ocelli. These are light-sensitive and can tell light from dark. Insects also have large compound eyes made up of many lenses that enable them to see in fine detail, even in the dark. They can see colours. They also see ultraviolet light, so they can use the Sun to navigate, even in cloudy weather. Insects hear by sensing sound vibrations with their eardrums. But most insects' ears are not found on their heads. The praying mantis, for example, has ears on its thorax between its hind legs.

Compound eye
Compound eyes are made of many lenses. Dragonflies, like the one shown here, have the largest eyes of any insect, with up to 30,000 lenses on each eye. Bees have about 5,000 lenses, while some ants have only nine.

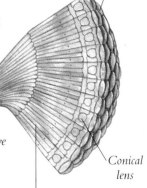

Hexagonal lens

Optic nerve

Conical lens

Nerve fibres

This diagram shows a cross-section through a compound eye. The surface is made up of tiny hexagonal (six-sided) lenses that fit together. Below are conical lenses that focus light on to nerve fibres, which carry signals to the brain via the optic nerve.

Horsefly

Compound eye

Seeing all around
An insect's compound eyes are found on either side of its head. Each eye builds up a slightly different view. The fly's brain probably interprets the images to get a sense of depth. The eyes extend around the insect's head, to allow it to see up, down and to the side all at once.

Through insects' eyes

Insect vision is very different from ours.
·Experts think each lens of an insect's
compound eye sees a small part of a scene.
This gives a mosaic-like view that is built
up into a bigger picture. These diagrams
compare how we see a moving insect and
what experts think an insect sees. An
insect can sense tiny movements our eyes
would hardly notice because they have
many more lenses that are affected. An
insect will be more sensitive than a human
to a bee taking off because a different set
of lenses is affected.

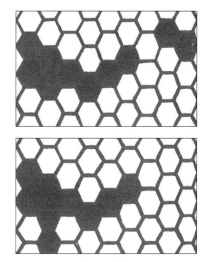

Human vision

Insect vision

Green lacewing

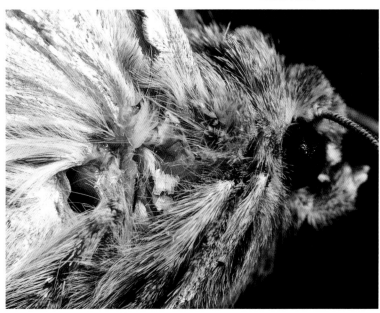

Insect ears

Insects' eardrums
are thin membranes
of skin. They vibrate
to sounds and send
signals to the brain.
Insects' ears are found in
various places. This green
lacewing hears with sensitive
hairs on its wings. A grasshopper has
ears on its abdomen. A cricket's ears are
on its front legs, just below the knee.

*This noctuid moth has its wings raised, clearly showing an ear hole on the side of
its thorax. The moth can hear the high-pitched squeak of a hunting bat and so
can escape its attacker. A bat's squeak is far too high for human ears to hear.*

TOUCH, TASTE AND SMELL

Drinker moth caterpillar

The body of this drinker moth caterpillar is covered with long, stiff hairs. The hairs are sensitive to sound waves in the air. If you clap your hands near a caterpillar, it will freeze, or defend itself by curling into a ball.

A N insect's main sense organs are its antennae. These projections are covered with tiny hairs that are sensitive to touch. But an insect's antennae can do much more than feel. Many insects use theirs to taste and smell, and even to hear as well. Smell is the most important sense for many insects. Some moths have long, feathery antennae that are sensitive to smell. Insects' antennae can also be sensitive to damp and heat. The human body louse uses its antennae to measure moisture in the air. It uses this to seek out the damp parts of humans where it makes its home. The antennae of blood-sucking bugs act as heat detectors. They are used to sense the body heat of the warm-blooded animals that the bugs feed on. Some insects have taste organs on their mouthparts. In many species, however, taste organs are located on the feet. Insects' bodies are covered with tiny sensitive hairs that detect vibrations in the air. The hairs help an insect to know which way is up and how fast it is flying.

Cockroach

Weevil

Antenna

Weevils like this have antennae branching from their snouts. The tips of the antennae are covered with sensitive hairs. The weevil's biting jaws are found on the end of the snout.

Cerci
Cockroaches and some crickets have special bristles called cerci on their abdomens. They can sense vibrations in the air caused by noise or movement. The slightest hint of danger will send the cockroach scuttling for cover.

Cerci

14

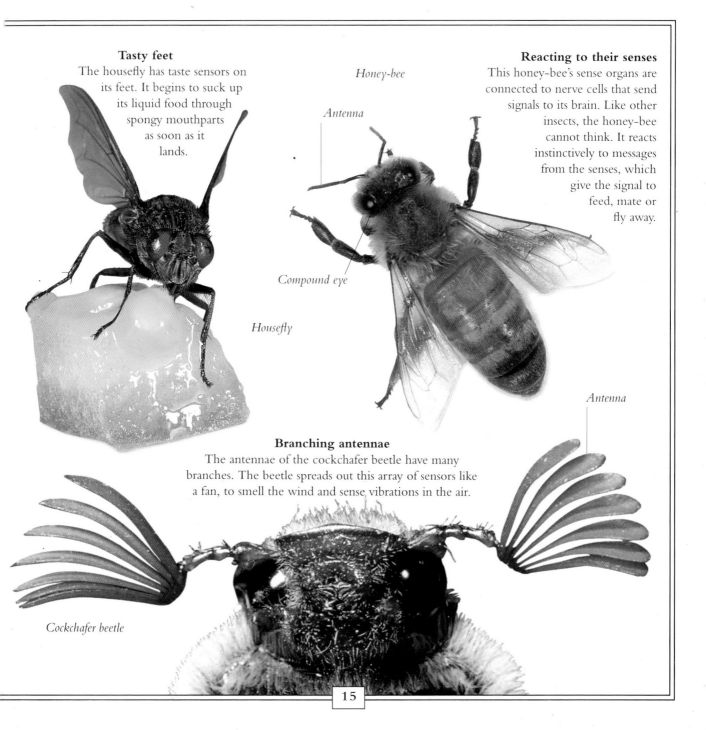

Tasty feet
The housefly has taste sensors on
its feet. It begins to suck up
its liquid food through
spongy mouthparts
as soon as it
lands.

Honey-bee

Antenna

Reacting to their senses
This honey-bee's sense organs are
connected to nerve cells that send
signals to its brain. Like other
insects, the honey-bee
cannot think. It reacts
instinctively to messages
from the senses, which
give the signal to
feed, mate or
fly away.

Compound eye

Housefly

Antenna

Branching antennae
The antennae of the cockchafer beetle have many
branches. The beetle spreads out this array of sensors like
a fan, to smell the wind and sense vibrations in the air.

Cockchafer beetle

REPRODUCTION

For any insect species to survive the adults must reproduce. Almost all insects do so by laying eggs. Most mate before the female lays her eggs. The sexes attract one another by sending special courtship signals. Many communicate by smell. Others use special sounds or visual signals. Male stag beetles, for example, have huge jaws shaped like stags' antlers. These antlers are used for fighting with other males in order to attract females. The female silk moth attracts the male by producing a special scent, called a pheromone. The male moth can smell the scent from several kilometres away and follows the trail of scent to find the female. After mating, many insects lay their eggs on a plant that will provide food for the young when they hatch out. Other species lay their eggs in water, on the soil or underground. After laying, most insects fly away and take no further care of their young.

This grasshopper is scraping its hind legs against its front wings. Rows of tiny pegs on its hind legs produce a harsh, chirping sound. Male grasshoppers make this noise to attract females and to warn other males away. Crickets make a similar sound by rubbing the rough edges of their wings together.

Stag beetles

Wrestling beetles

These male stag beetles are battling with each other for the chance to mate with a female. Only the males have these huge antler-like jaws, which they use for wrestling. They try to scoop their rival into the air with their antlers and throw the opponent to the ground. The antlers are too heavy to give a strong bite and are mainly for show. The males wrestle to advertise themselves to the females, when they are ready to mate. The females judge the males on how successful they are in competition.

Fast breeder

Some species of insect can reproduce without mating or even laying eggs. In spring, female aphids give birth to live young that are like miniature versions of themselves. The young aphids begin to feed immediately and can reproduce themselves when only a week old.

Mating beetles

A male and female lily beetle mate on a plant stem. In this species, mating takes several hours. A complicated courtship helps the female to make sure she has chosen a good mate. In other insects, mating takes only a few minutes.

Lily beetles

Caring for their young

Unlike most species of insect, female earwigs stay to guard their eggs and look after the young. The female earwig spends all winter guarding her eggs in an underground burrow. She licks the eggs clean and keeps them warm. When they hatch, she protects the young for a few days and feeds them with food from her stomach.

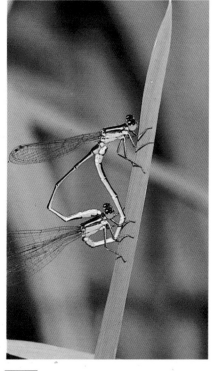

Heart shape

When a pair of damselflies mate, the male grasps the female's neck using claspers on the tip of his abdomen. The female then arches her own body up to collect sperm from the male to fertilize her eggs. As they mate, the damselflies make a heart-shape. They can fly along in this position while they are mating.

LIFE CYCLES

WHEN young insects hatch from their eggs they are ravenous eaters. They consume huge quantities of food and grow quickly. Like their parents, the bodies of young insects are protected by a tough exoskeleton. As they grow bigger, they must moult, or shed, this body case and grow a new one. As they moult, the young change shape, until they reach adulthood. This changing process is called metamorphosis. In some species of insect the changes happen gradually. For example, young grasshoppers and dragonflies are called nymphs and look like their parents, except that they have no wings. They change gradually as they grow, until they become mature. In other species the changes happen suddenly. Some young insects, called larvae, look nothing like their parents. Caterpillars and grubs are larvae. Their bodies must pass through a number of dramatic changes before they become adults.

Red admiral butterfly

Different kinds of insects spend varying amounts of time in their young and adult forms. This red admiral butterfly spends five weeks as a caterpillar, two weeks as a pupa and nine months as an adult. Mayflies, however, spend up to three years as nymphs, but only a few hours as adults.

Gradual change
These mantis nymphs (*left*) have just hatched. They look like their parents, but have no wings. As they get bigger they moult their outgrown skins. This gradual change is called incomplete metamorphosis. With each moult, the young insect becomes more like an adult (*right*). Gradually its wings develop from wing buds. Finally, it emerges from its last moult complete with wings and reproductive organs.

The elephant hawk moth begins life as an egg protected by a tough outer shell.

Eggs

Elephant hawk moth

Complete metamorphosis

Insects such as butterflies, moths, ants and bees change greatly during their life cycles. This is called complete metamorphosis. The life cycle of the elephant hawk moth shown here has four separate stages – egg, caterpillar, pupa and adult.

The egg hatches out a caterpillar. The caterpillar spends almost all its time feeding, stopping only to moult when its skin becomes too tight.

Caterpillar

The pupa splits and the adult moth emerges. It is ready to mate and lay eggs so the cycle can begin again.

Pupa

When the caterpillar is fully grown, it burrows into the ground and moults again to reveal a pupa.

Discarded nymph skin

Emerging adult

This cicada has just reached its adult form. It has moulted and wriggled clear of its old skin. Blood is pumped into the wing veins to expand the wings. Gradually the wings dry and harden, and the body gets darker in colour.

STUDYING LIFE CYCLES

Y OU can find out a lot about the life cycles of moths and butterflies by keeping caterpillars. First, prepare a home for the caterpillars using a cardboard box. Look for caterpillars on plants where you see half-eaten leaves and stems. You may find them hiding on the undersides of leaves. Make a careful note of the plant on which you found them and take some leaves with you. The caterpillars of small tortoiseshell butterflies, shown here, feed on stinging nettles. Use your field guide to identify the species you have found. Check the guide to see which plant they prefer. When picking up the caterpillars, try not to touch them directly with your fingers, as some species have hairs that will sting you. Pick them up with a paintbrush, or encourage them to climb on to a leaf. Carry them home in a collecting jar. At home, keep the cage out of direct sunlight, in a moist, cool place. Try to disturb the caterpillars as little as possible. Clean the cage out regularly and replace old leaves with fresh ones.

When the caterpillars become adult moths or butterflies it is time to let them go. Take the insects back to where you found them. Lift the lid off the cage and let them fly away.

You will need: scissors, cardboard box, strong sticky tape, muslin or netting, modelling clay, rubber gloves, fresh leaves, kitchen paper, collecting jar, ruler, pencil, notebook, field guide, coloured pencils.

Keeping caterpillars

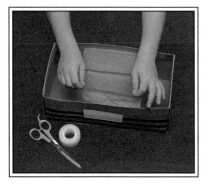

1 Cut holes in the sides of the box for windows. Using strong sticky tape, stick pieces of muslin or netting over the windows to cover them.

2 Now cut a large piece of muslin to make the cage lid. Weight the corners down with modelling clay to prevent the caterpillars escaping.

4 Put some damp kitchen paper in a corner to provide moisture. Carefully transfer your caterpillars from the collecting jar to the box. Cover with the lid. Check your caterpillars every day and replace the damp kitchen paper. Keep a diary of their life – how much do they eat and how big are they?

Remember to replace the leaves daily.

3 Wearing gloves, put fresh leaves in the cage. Make sure they are from a plant your caterpillars eat. Be sure to provide fresh leaves daily.

5 Watch how your caterpillars feed and move about. Record the dates when you see them moulting. How many times did they shed their skins?

Keep a chart of the life cycles of your insects.

June–July	caterpillar feeds
end July	
mid–August	pupa forms
	butterfly emerges

6 When it has finished growing, the caterpillar will change into a pupa, or chrysalis. It will attach itself to a twig and form a new skin.

7 Check your pupa every day and write down the date when you see the case splitting. How long did the insect spend as a pupa?

8 You will see a butterfly or moth struggle out of the old skin. The insect rests and pumps blood into its crumpled wings to straighten them before flying.

GETTING ABOUT

Egyptian grasshopper

Coxa

I NSECTS have six jointed legs for getting about. All insects' legs have the same basic structure. They are like hollow tubes with four parts. The leg is joined to the insect's thorax by the coxa. The femur and tibia are like your thigh and lower leg. The tarsus corresponds to your foot, and has two claws. All the sections meet at flexible joints. Muscles inside expand and contract (shorten) to allow the leg to move. Different species of insects move in various ways. The way they move can help you to identify them. The shape of an insect's legs help it to move around in a particular way. Insects that walk or run have long, thin legs. Those that move by jumping, such as crickets, have powerful back legs. A few species have strong front legs for digging. Some insects that live in water have wide, flat back legs covered with long hairs. The back legs move together and act like oars to row the insect through the water.

Femur

Joint

Tibia

Tarsus

When it is about to jump, the grasshopper gathers its long legs under its body. Leg muscles contract to straighten the leg, pushing the insect up.

Champion jumpers
Fleas are biting insects that feed on the blood of other animals. They have very powerful back legs to help them leap on to the bodies of much larger animals. These small insects are amazing jumpers. They can leap up to 30 cm in the air – 130 times their own height.

Flea

Springtails are not true insects, but they can also jump into the air. Unlike fleas, a springtail uses its forked tail to jump up rather than its legs. The tail is folded beneath the springtail's body. It flicks down against the ground, throwing the springtail forwards.

Housefly

Walking upside down

Houseflies and bluebottles have sucker pads and hooks on the soles of their feet. With the help of these pads and hooks they can crawl up smooth surfaces, such as walls and windows. They can even walk across the ceiling upside down.

Legs for digging

The mole cricket, like a real mole, has strong, shovel-like front legs. This unusual insect spends its life underground. Its front legs shovel earth aside as the cricket tunnels through the soil, looking for underground roots to eat. It has special scissor-like mouthparts for feeding.

A looper caterpillar, or inchworm, has two pairs of claspers on the end of its body. The caterpillar moves by bringing its claspers forward and arching its body into a loop. Then it stretches its front legs forward to move on.

Hawk moth caterpillar

Looper caterpillar

Clasper

Caterpillar movement

Caterpillars have three pairs of true legs and several pairs of false legs. You can see the true legs of this hawk moth caterpillar at the front of its long body. It has five pairs of false legs, called prolegs, at the rear. Prolegs are muscular projections like suckers. The caterpillar moves only one pair of legs at a time. This distributes its weight evenly along its body and helps it move over obstacles in its path.

FRESHWATER INSECTS

ANY kinds of insects are found in fresh water. Some live in large lakes, rivers or marshes. Others inhabit small ponds and shallow streams, or even live in puddles. Freshwater insects include many kinds of bugs and beetles. Most are predators, and feed on small creatures that have fallen into the water. Some insects spend only the early part of their lives in water. As nymphs or larvae, they take advantage of the abundant food supply found there to feed and grow. These young insects then go on to become adults that live on land. Just like other animals, freshwater insects need air to breathe. Some have developed tubes and gills to help them breathe underwater.

Walking on water

This pond skater has long, thin legs and a lightweight body to help it walk on water. It spreads out its legs and skims along on the surface. A skin on the surface of the water prevents the insect from sinking. The pond skater moves by rowing with its back and middle legs.

The water-boatman or backswimmer swims upside down below the water's surface. Its hind legs are like oars and row the insect along. When it senses the vibrations made by other creatures it rushes over to stab the victim and drink its juices.

Breathing underwater

The water scorpion has a tube at the end of its abdomen. When it needs to breathe, it raises the tube to the surface like a snorkel. The tube is fringed with hairs so water cannot run down inside.

The water beetle traps a bubble of air under its wings before it dives. It uses this as an air supply to breathe from as it swims underwater, just like a diver using an oxygen tank. The beetle has to swim strongly so that it does not bob back up to the surface.

Watery home

The larva of the caddis fly lives underwater. Its spins itself a protective case of silk. To keep this home well hidden, the larva attaches small sticks and stones from the pond or stream-bed to the outside of the case.

From water to land

The damselfly spends its early life as a nymph living underwater. Three feathery gills on the tip of its abdomen filter oxygen from the water to breathe. When it is fully grown, the nymph (*left*) uses a plant stem to climb out of the water. Its skin splits and a young adult emerges. The young adult damselfly (*right*) rests after it emerges. The abdomen and thorax lengthen as the insect stretches out its crumpled wings. When its wings are dry, the damselfly takes to the air for the first time.

WATCHING POND INSECTS

You will need: gardening gloves, trowel, washing-up bowl, gravel, water plants, large stones, watering can.

Y OU will find many kinds of freshwater insects in your local pond or stream. Spring and summer are good times to look for them, because the young insects turn into adults at these times. You could make a small pool for insects in your garden or perhaps at school. Ask a responsible adult if you can dig the pool. To catch water insects for study you will need a net, which you can make quite easily yourself. When you go to catch insects at the pond, take an adult with you for safety. As you approach the water, move quietly or you will disturb the wildlife. Never run near water – you could easily trip and fall in. Different insects live in various places in the pond or stream. Some live near the surface, others swim near the bottom. Lift up stones and pebbles to find the creatures that lurk on the underside. Record the date, time and place where you found the insects. Try visiting different ponds and streams to see if the species you find vary.

Make an insect pool

1 Wear gloves when making your insect pool. Dig a hollow in the ground with the trowel. The hole should be big enough to fit an old washing-up bowl inside.

2 Place the bowl in the hollow and press it down firmly. Spread gravel on the bottom and put in the water plants. Place stones around the edge of the bowl and inside it.

3 Then fill the bowl with water, using a watering can. Your pool is now finished. Insects and other animal life will soon be attracted to the pool.

Make a pond net

1 Begin your net by threading wire in and out through the top of a thin sock. You may need pliers to bend the wire into a circle.

2 Use pliers to twist the ends of the wire together to make the net secure. Now position the net at the end of a long pole.

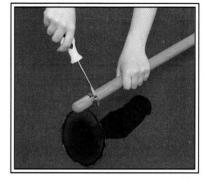

3 Thread the jubilee clip over the pole and push the twisted wires under the clip. Tighten the clip using a screwdriver.

4 At the pond, capture insects by sweeping your net gently through the water. Lightly tap the stems of plants to knock other insects into your net.

MATERIALS

You will need: wire, thin sock, pliers, long pole or broom handle, jubilee clip, screwdriver, jug, empty plastic ice-cream container, magnifying glass, field guide.

5 Empty a jug of pond water into a plastic container. Now carefully empty your net into it, too. Study the insects and other creatures you have caught. Your field guide will help you identify them. Gently tip the water and creatures back into the pond when you have finished looking at your finds.

INSECTS IN FLIGHT

Like all flies, crane-flies have only one pair of wings. Instead of a pair of rear wings they have tiny balancing organs, called halteres. Halteres look like little knobs on stalks. They help the crane-fly to manoeuvre in the air.

INSECTS have been flying around on Earth for a staggering 350 million years. The power of flight has helped insects to escape their enemies and travel to new areas to find food. Most flying insects have two pairs of wings, but different species have various wing shapes. The way they fly is varied too, so you can identify some insects by their flight pattern. Butterflies, for example, have two pairs of large, triangular wings. They flap their wings slowly and flutter here and there. Bees and wasps have narrower wings and fly a straighter path. Midges have narrow, transparent wings with long fringes. These tiny insects seem to dance as they hover in the air. The buzzing sound many insects make as they fly is the sound of their wings beating. The wings are made of the same hard material, called chitin, as the rest of the body case. Veins fan out across the fragile wings to strengthen them. Many insects fold their wings away when they are on the ground. They must protect their wings because, if damaged or broken, they will not grow back.

Linked wings

The front and rear wings of butterflies overlap to make a single surface. Butterflies are strong fliers although they flap their wings quite slowly.

Moths have special bristles on their hind wings that act as hooks to link their wings together. In flight, the bristle locks under a tiny catch on the front wing. You can see the dark bristle under the light-coloured catch in this picture of a buff ermine moth's wing.

Beetles' wings at rest

Cockchafers and other beetles have hard wing cases instead of true front wings. When the insect is resting, the hard cases cover and protect the thin flying wings underneath.

Beetles' wings in flight

As a cockchafer prepares for take-off, its wing cases lift up and the back wings unfold. In flight, the wing cases are spread out and help to keep the insect airborne.

How insects fly

Insects have no muscles in their wings to beat them up and down. Instead the wings are hinged to the insect's thorax. They move up and down as the thorax changes shape. As the roof of the thorax is pulled down, the wings flick up. As the ends of the thorax are pulled in, the wings flick down.

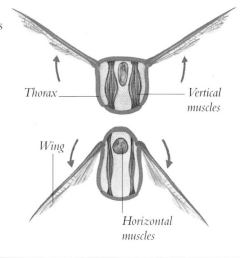

Thorax — *Vertical muscles*

Wing — *Horizontal muscles*

MIGRATION

Most insects spend their whole lives in one place. Others make long seasonal journeys called migrations. They fly hundreds of kilometres to a different area and back again each year. Other kinds of animals go on migrations too, including many species of birds. Most insects migrate to avoid the cold of winter, or to find food for themselves or for their young. A few insects, such as bogong moths in Australia, migrate to cooler areas to avoid the fierce summer heat and drought. Over 200 types of butterflies and many kinds of moths migrate. These insects find their way by instinct, following the Sun's path as it moves across the sky. Most fly and feed by day, resting at night. Migrating insects have travelled the same routes for centuries. They come and go at regular times each year. Other insects make irregular journeys, called irruptions. These flights are triggered by hunger when food is scarce.

Painted lady butterfly

The painted lady is found in many parts of the world. These colourful butterflies make long annual migrations. Painted ladies that hatch out in Africa travel north to Europe in the spring. Their offspring travel back again in the autumn.

Locust pest
Locusts live in the dry parts of Africa, Australia and the Americas. The locusts usually live alone, but after the rain they breed in large numbers and the young mass together. When food becomes scarce, they take off in huge swarms containing billions of flying locusts. They can cover an area of 1,000 square km in search of food.

Desert locust

A swarm of locusts
These locusts are descending on a farmer's field. In minutes they will have eaten all the crop before moving on again. A swarm may stay together for several years.

Long-distance champion

The monarch butterfly is the champion migrant among insects. Butterflies starting out from southern hibernation sites in the spring lay eggs as they go. Their offspring reach Canada, thereby completing the round trip of over 6,000 km begun by their ancestors.

In September the monarch butterflies head southwards across the United States. Their flight looks fluttery and aimless, but monarchs fly at a steady 10 km per hour and cover up to 130 km each day.

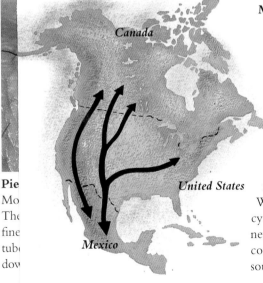

Monarch butterfly

Migration map

This map shows the path of the monarch butterflies. In late autumn they arrive in Florida on the east coast of the United States, California on the west coast or Mexico. There, they gather on trees to pass the winter. In March, the butterflies begin the journey north, lay their eggs and die. When the eggs hatch out the cycle begins again, with the newly-hatched butterflies either continuing north or returning south, depending on the season.

Canada

United States

Mexico

Pie
Mo
The
fine
tub
dow

FACT BOX

• Desert locusts are usually green in colour. But when they swarm, their colour changes. As they cluster together, they touch one another. This triggers a chemical reaction that changes their body colour to yellow, red or black.

• In Australia, bogong moth caterpillars hatch out in the south-east lowlands in spring. In summer the lowlands heat up, and the adult moths migrate to the Australian Alps where it is cooler. There, they hibernate in crevices in the rocks until the autumn, when it is cool enough to fly back to the lowlands.

SIGNS OF FEEDING

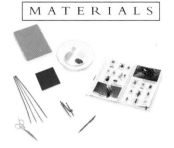

You will need: stiff card, compasess and pencil, scissors, four garden sticks, small samples of food (such as jam, meat, cheese and fruit), notebook, field guide.

As insects eat, they leave damaged plants and other signs of feeding. Sometimes these signs are easier to spot than the insects themselves. Choose a small area, such as a fallen log, a bush or a group of plants. Hundreds of insects will be near, but most are small and wary. Look along cracks or crevices in the bark where insects may be hiding. Look under leaves and flowers. Any curled-up leaves may be an insect's home. Most plant-eating insects prefer one particular food, and may eat only a part of that food plant. You will find butterflies and bees on flowers, drinking the sugary nectar. Caterpillars gather and feed on leaves. Aphids such as greenfly live on the stems of plants and suck out the sap. Find out if insects prefer certain foods by watching which ones visit different food samples.

Food samples

1 Make four circles on the sheet of card with the compasses and pencil. Cut out the circles. Using the point of the compasses or a sharp pencil, make holes in the centres of the circles. Push the sticks through the holes so the card circles stay in position.

2 Outside, plant the sticks in the ground. Push the food samples on the sticks, so they rest on the circles. Do more insects visit some food samples than others? Are there more insects around at different times of day?

Finding signs of feeding

1 In spring and summer, check the leaves of plants and trees for signs of insects feeding. Use your magnifying glass to take a closer look and make notes of what you find.

2 Some insect larvae strip plants bare and nibble through stems. Others leave great ragged holes in leaves. Your field guide may tell you which larvae are eating the plants.

3 Aphids and other bugs leave brown or yellow lines on crops when they suck out the sap. Look out for greenfly and other aphids on the stems of roses, too.

MATERIALS

You will need: magnifying glass, pen or pencil, notebook, field guide.

5 Look for white or brown markings on fresh leaves. Hold a leaf up to the light. Can you see any larvae inside? These are leaf miners. They tunnel inside the leaf and eat the tissue. Some eat patches of leaf, while others leave a winding trail.

4 Many insects tunnel into wood. The lines on this tree trunk were made by bark beetle larvae as they chewed their way between the bark and the underlying wood.

INSECTS AND PLANTS

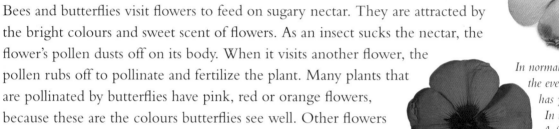

MANY insects depend on plants, but many plants depend on insects, too. To make seeds, a plant must be fertilized by pollen from the same plant or another of its species. Insects help by carrying pollen from one plant to another. Bees and butterflies visit flowers to feed on sugary nectar. They are attracted by the bright colours and sweet scent of flowers. As an insect sucks the nectar, the flower's pollen dusts off on its body. When it visits another flower, the pollen rubs off to pollinate and fertilize the plant. Many plants that are pollinated by butterflies have pink, red or orange flowers, because these are the colours butterflies see well. Other flowers have special markings, called nectar guides, radiating from the base of the petals. Some show up only in ultraviolet light. The eyes of insects such as bees are sensitive to ultraviolet light, and they follow these guides to find the sweet nectar in the middle of the flower.

In normal light (above) the evening primrose has yellow petals. In ultraviolet light (left) the flower's nectar guides show up as a dark patch, to lead the insect to the nectar in the flower's centre.

Bumble-bee

Insect pollination
This bumble-bee is visiting a flower to feed on nectar. It sucks up the nectar with its long tongue and gathers pollen to feed the larvae in its nest. As the bee reaches into the centre of the flower, pollen from the flower's stamens (male parts) rubs off on the bee's hairy body, coating it with dusty pollen grains.

After visiting a flower, the bee combs the pollen grains from its body with its legs. It gathers the pollen into tiny pockets on its back legs. But some grains stay on the bee's body. When it visits the next flower, the pollen rubs off on to the stigmas (female parts) of the second flower. There they fertilize the plant.

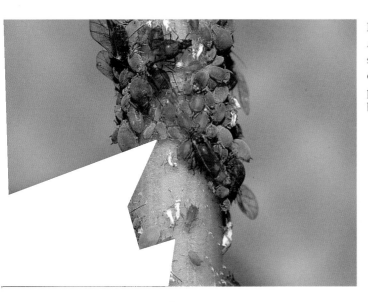

Plants under attack

A group of blackfly and their young feed on a plant stem. When their mouthparts pierce the outer layer of the stem, they suck up the sap. Sap is made by plants in their leaves as food for themselves, but it is also a very good food for insects.

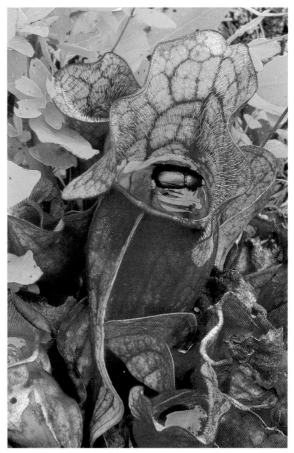

Pitcher plants have jug-shaped traps with slippery sides. When an insect lands on the rim, it slips down inside the pitcher. It falls into the liquid at the bottom and drowns. Then the plant absorbs the nutrients from the insect's body.

Venus fly-trap

Plants bite back

The Venus fly-trap has paired leaves with sharp spines on the edges. When a fly lands on the open leaves, they snap shut and the fly is trapped inside. Chemicals in the leaves slowly dissolve the insect's body. The plant then absorbs nutritious minerals from the fly.

WOODLAND INSECTS

M A T E R I A L S

You will need:
old white sheet, collecting jar,
paintbrush, magnifying glass,
field guide, notebook, pen or
pencil, coloured pencils.

Woods are great places to go insect-watching – but remember to always take an adult with you. Woodland trees offer plenty of food and shelter from strong winds and weather and so are an ideal habitat for literally millions of insects. The number of insects you find may depend on the season – in spring, wild flowers bloom under the trees, attracting insects, and in summer the woods offer insects sunny clearings and cool shade. Mixed woodland with several different types of trees is best for insects. The larger trees of the wood, such as oak and beech, are home to hundreds of different species. Choose a large tree and look at all the different parts – leaves, twigs, fruits or blossoms, bark and roots – or make a survey of all the insects you can find on a single branch. Make a tree trap to catch insects active at night.

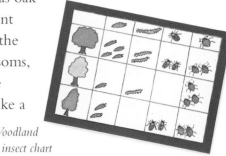

Woodland insect chart

Studying life on a branch

1 Spread out the white sheet below a branch. Shake the branch to dislodge the insects on to the sheet. If the branch is high, tap it with a stick. Be sure not to damage the tree.

2 Sweep the insects that drop on to the sheet into collecting jars for study. Use a paintbrush to carefully transfer the insects without harming them.

3 Use a field guide to identify the insects. Try surveying another tree. Make a chart, as above, to show the different species found on the different trees.

Make a tree trap

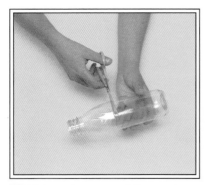

1 Using scissors, carefully cut the plastic bottle in half. You may need an adult to help with this.

2 Turn the neck of the bottle round and push it inside the bottom half. Now tape the two halves together using sticky tape.

3 Cut a long piece of string. Loop the string round the open end of the trap and secure with a knot. Bait the trap with a small piece of ham.

4 Tie the trap along a tree branch, or hanging down underneath the branch. Leave the trap out overnight.

5 Go back the next morning to check your trap. Identify the insects you have caught with your field guide. Record your findings in a notebook. Release the insects when you have identified them.

COLOUR AND CAMOUFLAGE

INSECTS have many enemies in the natural world. They are eaten by a wide range of animals, such as bats, frogs, lizards, birds, mammals and even other insects. Most adult insects try to escape predators by flying away. But it is even better not to be seen at all. Many insects have special colours and patterns on their bodies to help them blend in with the natural world, so predators do not see them. Some mimic a natural feature, such as a thorn or a twig. Predators hunting for food do not focus on their surroundings. So an insect that looks like a twig will not be noticed. These natural disguises are known as camouflage. Different kinds of insects imitate a wide range of natural objects. Some look like leaves, seeds or flowers. Others resemble moss, bark or lichen, sticks or stones. Camouflage helps insects to hide from danger. It also helps insects that hunt other creatures to creep up unnoticed on their prey.

The toad grasshopper lives in dry parts of Africa. Its flattened body shape and colour help it to blend in with the bare rocks and stones of its habitat.

Hidden hunter

Can you see the mantis in this picture? The insect is so well camouflaged that it is very hard to spot against the bark. Mantises are predators, hunting other animals for food. This disguise will help it approach its prey unseen.

FACT BOX

• Red, yellow and orange colours in insects are usually caused by chemicals left over from the insects' food.

• Caterpillars are particularly at risk from predators because they cannot fly. Camouflage helps many to keep well hidden.

Changing camouflage

Peppered moths usually have whitish wings and bodies with black speckles. This colour helps them blend in well with tree bark. During the 1800s, however, many trees near towns were blackened with factory soot. Over the years, a new, darker variety of peppered moth became common in towns, because it blended in much better with its new surroundings.

This is the pupa of a swallowtail butterfly. In its dormant state, it is very vulnerable to attack from predators. The pupa's skin is coloured to blend in with the plant stem to which it is attached. Camouflage keeps the insect safe while it changes into a butterfly.

Leaf mimic

This bush cricket is from West Africa. The colour, shape and patterns of its wings help it to mimic the leaves of its rainforest home. It even has lines on its wings that look like leaf veins. This makes it very difficult to recognize amongst the leaves.

Acting like a twig

This waved umber moth caterpillar is camouflaged to look like a twig. It completes its disguise by holding its body at an angle on the branch. In this position it looks just like a twig branching from the main stem. When danger threatens the caterpillar keeps absolutely still.

WARNINGS AND DISGUISES

Bullseye moth

Some kinds of insects are armed with poisonous stings or bites. Others have poisonous or foul-tasting fluids in their bodies. These sorts of insects do not need to hide from predators. Instead, they make themselves quite obvious. Their bodies have special warning colours that are used and understood throughout the animal world. Warning colours are often bright red or yellow, with black spots or stripes. These patterns are easily seen by predators. Ladybirds, for example, have bright red wing cases patterned with black spots. If an animal tries to eat a ladybird it produces a foul-tasting fluid. Predators that have tried once will not try again – they recognize the colours and avoid ladybirds in future. Other insects take advantage of warning colours. They are not poisonous and cannot sting or bite, but their bodies have the same colours as a poisonous species. Fungus beetles are red with black spots, just like ladybirds. They are harmless, but because they look like ladybirds, predators leave them alone.

Eyespots
The bullseye moth has two large eyespots on its hind wings. When its wings are open, the spots look like the eyes of a much larger animal and so scare off a predator.

Warning colours
Wasps are dangerous insects. Female wasps are armed with a venomous sting that can pump poison into an attacker. Animals that eat insects know to avoid wasps, or soon learn by experience. The common wasp's body has yellow-and-black striped warning colours.

Common wasp

Hoverfly

Harmless mimic
This insect looks very like a wasp with its vivid yellow-and-black striped body, but in fact it is a hoverfly. Its warning colours mean that most animals, including humans, leave them well alone. Hoverflies are harmless, however, and have no sting. This defensive strategy is called mimicry.

Startle colours

While resting on a tree trunk with its wings closed (*left*), this moth is well camouflaged against the bark. If, however, a predator comes too close, it spreads out its front wings (*right*) to show its colourful hind wings. The bright flash of colour may startle the hunter and distract it for a moment. Meanwhile the moth flits to another perch, and closes its wings, becoming invisible again. The use of colours to startle predators is quite common among insects and is a useful means of stalling a potential attacker for a few precious seconds.

A threatening display

Wetas are large, rare insects from New Zealand. When startled, the weta raises its large, toothed hind legs above its head. Many creatures will be frightened away by this menacing sight. Those that are not scared off may receive a painful kick from the weta's powerful legs.

The chrysalis of the black hairstreak butterfly looks just like a bird dropping. Most birds will not taste their own droppings when searching for food, so the hairstreak is left alone. Looking like something inedible is a clever way to hide from predators, especially if you cannot fly.

NATURE DETECTIVES

You will need: two cardboard boxes, scissors, light and dark green paper, sticky tape, paintbrush, privet or ivy leaves, kitchen paper, muslin, modelling clay, notebook, coloured pencils.

FOR an insect's camouflage to work it must be hiding against the right colour background. If it moves to a different place it may become obvious to predators and open to attack. Some insects, however, can change their body colour to match their surroundings so that they remain hidden almost anywhere. Stick insects can do this and are experts at disguise. They have long, slender bodies and stick-like legs, which makes them very hard to see among twigs and leaves. Buy some stick insects from a pet shop and try the test below to find out more about camouflage. Remember to ask the pet shop what your stick insects like to eat. Stick insects are easy to rear and look after at home. You could also look outside for ways young insects disguise themselves.

During the day stick insects stay very still. They usually move and feed at night.

Camouflage test

1 Cut pieces of coloured paper to line the insides of the cardboard boxes. Make one box light green and the other dark green. Attach the paper with sticky tape.

2 Transfer your stick insects with a paintbrush to the light green box. Add leaves and damp kitchen paper. Cover with muslin weighted at the corners with modelling clay.

3 Leave the box in a light place for a day. Look at the insects and record their colour with coloured pencils. Transfer the insects to the dark green box. After another day, check them to see if they have changed colour.

Rearing stick insects

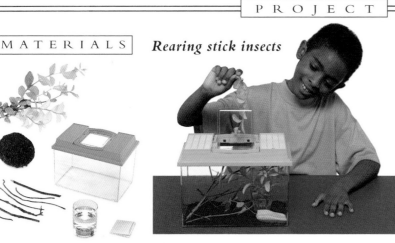

You will need: small tank or large empty sweet jar, earth, privet or ivy leaves, glass of water, sticks, kitchen paper.

1 Keep your stick insects in a tank or a large sweet jar. The tank must have a tight-fitting lid with small air holes. Put a layer of earth in the bottom. Add some privet or ivy leaves in a glass of water, and some sticks.

2 Put some wet kitchen paper in a corner so the insects have enough moisture. Remember to replace this regularly. Ask the pet shop if your insects need anything else.

Insect home
Would you recognize this object as an insect's home? It is called a robin's pincushion and is in fact a gall – a swelling on a plant caused by an insect. This gall is formed when a tiny wasp lays its eggs on a rose bud. It contains many developing wasp larvae.

Hiding place
The nymph of the froghopper bug protects itself from hungry enemies by producing a blob of unpleasant-looking foam. The foam is called cuckoo-spit and is a common sight on plants. The nymph is usually hidden in the middle of the foam, where it can safely suck the plant's sap.

ATTACK AND DEFENCE

*Puss moth
caterpillar*

INSECTS that feed on living animals, including other insects, must be able to hunt and catch their prey. They must also have a way of overpowering their victims. Insects can see movement well, but they have more difficulty in spotting things that are completely still. Some hunting insects take advantage of this weakness. They lie in wait for another insect to pass by, and then spring out if it comes close. Many predators that hunt by stealth are well camouflaged – if they keep still their victims will not notice them until it is too late. Some insects catch their prey by seizing them in powerful jaws. Others use sharp spines on their legs. Some insects are armed with stings or poisons. These may be used either for attack or for defence. Wasps and ants have venomous stings or bites that can be used to overcome prey. Insects also defend themselves in various 'ways, often wounding their attackers. Some have poisonous bodies or taste horrible, others have sharp spines or can squirt stinging liquid.

The puss moth caterpillar has several lines of defence if it is threatened. The caterpillar rears up its head to reveal a bright red frill and large black eyespots. At the same time it waves its tail, which looks like a snake's forked tongue. If this does not scare the enemy away, it squirts stinging acid from a gland in its thorax.

FACT BOX

• The pupa of the South African leaf beetle contains a powerful poison, which South African bushmen tip their arrows with for hunting. If an arrow so much as grazes an animal's body, it is sure to die.

• When it is threatened, the bombardier beetle mixes two chemicals in its abdomen. This produces a small explosion, and a blast of hot gases shoots from the insect's rear.

Prickly meal
Caterpillars make a juicy meal for predators such as birds and lizards. To protect against predators, some species, such as this emperor moth caterpillar, have sharp, spiny bristles on their bodies. The spines prick a predator's mouth and make it drop the insect.

*Emperor moth
caterpillar*

Stinging weapons

This sand-tailed digger wasp is holding a weevil it has paralysed with its sting. Many wasp species use poison to capture prey. The sting is a modified ovipositor, the organ used to lay eggs. This means only female wasps can sting.

This is a close-up of a wasp's sting. The barbed spine is connected to a poison sac that pumps poison into the wound.

Sting

Poison sac

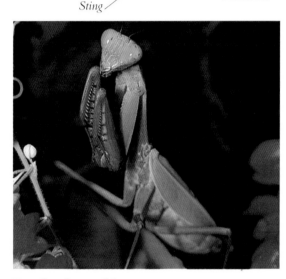

Hunting by stealth

The praying mantis is camouflaged to look like the leaves or flowers of the plant it lives on. It hunts by keeping very still. Only its eyes swivel as it scans for victims. When an insect lands nearby, the mantis darts forward with lightning speed. It grabs the victim in its front legs. The sharp spines on its legs keep a firm grip while the mantis eats its prey alive.

Acid attack

These green tree ants are attacking a caterpillar. Ants have strong jaws to bite their enemies and prey. Some also squirt stinging acid into the wounds they make. This group of ants work together to kill the caterpillar and carry off its body.

WASPS AND BEES

ASPS and bees belong to the insect order called Hymenoptera. There are thousands of different kinds and they live in almost every part of the world. Many kinds of wasps and bees have yellow and black warning colours and many have stings. Most are solitary insects and live alone. But bumble-bees, honey-bees and social wasps live in large groups called colonies. Honey-bees may nest in a tree or a beehive. The colony is made up of a single queen, hundreds of drones and thousands of worker bees. The queen is the largest bee. Her task is to lay eggs. The drones are male. Their job is to mate with the queen, so the eggs can be laid. The workers are undeveloped female bees. They do all the work of the colony, including building and repairing the nest. They also gather nectar and pollen from flowers and bring it to the nest. The workers feed the larvae, or grubs, with pollen and honey. In a few weeks, the grubs become pupae and then adult bees. All bees and wasps go through complete metamorphosis to become adults.

A bee keeper removes part of a hive to gather honey. People have collected honey for many centuries. Bees are kept in hives so that the honey can be harvested easily. The bee keeper's heavy clothing protects him from stings.

Bees make honey to feed their larvae and to keep themselves alive in the winter.

Inside the hive
Worker honey-bees build six-sided cells from beeswax. They make them on sheets called combs. Some cells are filled with honey and pollen. Others contain eggs or developing larvae.

Bee dance
When a bee finds a source of nectar, it tells the other bees by doing a special dance like a figure of eight. The angle of the bee's body on the comb tells the other bees the direction of the food. If the nectar is close, the bee waggles its body energetically.

You may see wasps on a fence post gathering wood. This is because common wasps build nests of paper. They make the paper by chewing up wood fibres. If you notice a wooden post with tiny parallel grooves, these may have been made by a wasp collecting fibres.

FACT BOX

• In the 1800s it was fashionable for ladies to have narrow waists like wasps. The "wasp waist" was created by a tight corset, which was very uncomfortable.

• Male orchid bees from South America make a perfume from orchid flowers to attract females.

• Paintings in tombs from Egypt prove that ancient Egyptians kept bees for honey 2,500 years ago.

Inside a wasps' nest

This wasps' nest has been cut away so you can see inside. The outer layers of paper protect the cells in the middle. A single larva develops in each cell. Worker wasps build combs, clean the cells, feed the larvae and forage for food. They feed their grubs chewed-up insects.

Wasp's nest

Laying eggs

The giant wood wasp is not a true wasp and has no sting. The female lays her eggs under the bark of a pine tree. She drills through the bark with her ovipositor, the organ on the tip of her abdomen. Some kinds of wasps lay their eggs on or in other insects. When they hatch out, their grubs feed on the host insects and eventually kill them.

Wood wasp

ANTS AND TERMITES

ANTS and termites belong to different insect orders, yet they behave in similar ways. They are social insects and live in colonies. A colony may contain thousands or even millions of insects, each with its own role. Queen ants (there may be several in one nest) lay eggs. As in a beehive, the tasks of a colony are carried out by undeveloped females called workers. Some ant species have workers with large jaws, called soldiers, who guard and defend the nest. In summer, winged male and female ants fly out of the nests to mate and the females start new colonies. In termite colonies the queen has a king who stays by her side to fertilize the eggs. Workers can be either male or female.

Wood ant

A wood ant smells the air with its antennae. Ants cannot see very well, so touch and smell are vital senses. When ants meet, they touch antennae. Intruders with the wrong scent are driven away.

Black worker ants carry their pupae to safety when their nest has been disturbed. In the nest, they tend and feed the larvae, and clean them with their saliva.

Working together
In South America leaf-cutter ants work together to bite off large pieces of leaf. Their powerful jaws slice through the leaves like scissors. Then each ant carries a piece bigger than itself back to their nest, where they are used to grow a special fungus that the ants feed on.

Leaf-cutter ants

Queen's chamber

Rubbish

Eggs

Pupae

Workers

Young ants hatch

Worker and larvae

Inside an ants' nest
Ants' nests are usually underground. A nest has many chambers, or rooms, and passages. Different chambers contain the eggs, larvae, pupae and the queen. Other chambers are used to store food and rubbish. Worker ants alter the temperature of the nest by opening or closing passages.

Inside a termites' nest

A termite colony has chambers for the young and a large chamber for the queen, where eggs are produced. The queen may lay an egg every three seconds, and live for 15 years! The workers and soldier termites are blind. The queen communicates with them using special scents called pheromones. Some scents tell the workers that it is time to gather fungus as food, whilst others send the message that they must tend the young.

Air escapes

Ventilation shaft

Air enters

Larval chambers

Queen's chamber

Fungus chambers

Towering home

A few kinds of termites build spectacular termite mounds like this one in Africa. Underground nests have tall towers that rise high into the air. Above the chambers, these towers have ventilation shafts built of soil that circulate the air inside the nest and keep it cool.

Workers and soldiers

Termites live in warm grasslands all over the world. They feed mostly on plant matter and do a valuable job in recycling nutrients back into the soil. Worker termites are wingless. Soldier termites have large, armoured heads. Their job is to defend the nest.

INVESTIGATING ANTS

M A T E R I A L S

*You will need:
gardening gloves, peeled
ripe fruit, piece of paper,
magnifying glass.*

A N ant colony is like an underground city. Each citizen has their own job to do. The workers scurry to and fro, scouting for food and bringing it to the nest. If the nest is disturbed, the ants swarm out to defend their home. The key to the smooth running of the colony is good communication. Ants communicate mainly through touch and by using powerful pheromones. Different scents tell the worker ants that the larvae must be cleaned or fed, or that the nest should be repaired. When an ant finds a new source of food, it hurries back to the nest. As it runs along, it presses its body on the ground. This leaves a trail of scent which the other ants can follow to reach the food. You can watch ants communicating in this way by putting down food near an ant's nest. An even better way to study ants, however, is to start a colony. If you establish an ant colony in a glass jar covered in dark paper, and uncover it after a few days, you will be able to see the ant tunnels.

Watching ant trails

1 Wearing gloves, find a trail of ants. Follow the trail to find out where the ants are going. Does the trail lead to food? Rub out part of the trail and see what happens.

2 Now put fruit down on a piece of paper near the trail of ants. The paper will make it easier to see the ants. When the workers find the fruit, watch to see what happens.

3 Once an ant has laid a scent trail to the fruit, others follow. Move the fruit to another part of the paper. What happens next? Do the ants go straight to the new food site?

Make an ant home

1 Cut a piece of paper large enough to fit around the jar. Fix the paper in position with sticky tape.

2 Wearing gloves, use the trowel to fill the jar with earth until it is almost full. Add some leaves on top.

3 Capture some garden ants using a paintbrush and collecting jar. Transfer the ants to their new home.

M A T E R I A L S

You will need: large jar, dark-coloured paper, scissors, sticky tape, gardening gloves, trowel, earth, leaves, paintbrush, collecting jar, ripe fruit or jam, kitchen paper, muslin.

If you have caught a queen, you may see the workers tending eggs or larvae in special chambers.

4 Feed your ants with a piece of ripe fruit or jam. Some damp kitchen paper will provide moisture. Feed your ants daily and refresh the leaves and moist paper regularly.

5 Cover the top of the jar with a piece of muslin, so the ants cannot escape. Secure the muslin with sticky tape. Keep your ant home in a cool place.

6 In a few days, lift the paper to see your ant home. There will now be winding tunnels, built against the sides of the jar.

BEAUTIFUL BUTTERFLIES

BUTTERFLIES' wings are covered with thousands of tiny overlapping scales. The scales are brightly coloured, or reflect the light so that they shine with colour. The natural beauty of butterflies has made them a target for collectors for hundreds of years. The survival of some species is now threatened by the activities of collectors. Other butterflies are in danger of dying out because their habitats are being cleared or drained for farmland or to make room to build houses. Some kinds of butterflies have wings with warning colours. These markings tell predators that they are poisonous to eat. Poisonous butterflies of different species in the same area sometimes look identical. This strengthens the warning message sent to predators. Other butterflies are harmless, but mimic poisonous species. Their colours fool predators into avoiding them as well. All butterflies go through complete metamorphosis, changing from caterpillars to pupae before they become adults.

A close-up of the overlapping scales on a morpho butterfly's wing. Light bouncing off tiny ridges on the scales produces an intense blue colour.

Painted lady butterfly

Swallowtail butterfly

At rest
Like most butterflies, this painted lady rests with its wings closed. The undersides of its colourful wings are dull. This helps to conceal it from a predator when it lands.

Club antennae
It is quite difficult to tell a butterfly from a moth, but a good guide is to look at the antennae. This swallowtail has the antennae with clubbed tips like most butterflies. Moths' antennae vary a great deal, but most are straight or feathered.

Male and female

The colours and patterns of butterflies' wings help them to recognize others of their species. Males and females often have slightly different markings, so they can recognize each other for mating. These are both orange tip butterflies. Only the male (*right*) has orange tips on his wings. Females (*left*) have grey wing tips.

Viceroy butterfly

Monarch butterfly

Winter sleep

Butterflies cannot stay active in cold weather. Some find a safe, sheltered spot for the winter and hibernate. This peacock butterfly is hibernating on a tree trunk. Its dark undersides help to conceal it in dark corners.

Copy-cat colours

These two butterflies look very similar, but they belong to two different species. Monarch butterflies are poisonous – their caterpillars feed on poisonous milkweed plants and can store the poison in their bodies. The viceroy butterfly is harmless, but looks just like the monarch. Birds and reptiles see the warning colours and do not try to eat it.

BUTTERFLY GARDEN

MATERIALS

You will need: gardening gloves, window box or large tub, earth or compost, packet of wild flower seeds, watering can, notebook, pencil, field guide.

THE best way to attract butterflies is to plant a butterfly garden. You could make your garden in a small space, such as a window box, or it could be larger with room for many different plants and herbs. Choose plants that bloom at various times of the year. Different plants attract butterflies and their larvae. Caterpillars feed on particular leafy plants, such as grasses, thistles and nettles. Adult moths and butterflies gather on plants with nectar-bearing flowers. Wallflowers, buddleia, golden rod, candytuft, ice-plants and hebes will attract adult insects. Moths are attracted to honeysuckle because the flowers release a sweet smell at night. You will have to ask a responsible adult if you can grow these plants. Try not to use insecticides in your part of the garden. In a small space such as a window box you could plant verbena, phlox and alyssum. You could also grow herbs such as marjoram and thyme. When studying butterflies, keep very still. Do not let your shadow fall on the insects or you will frighten them.

Plant a window box

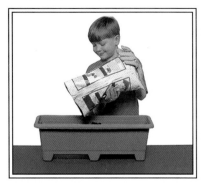

1 Wearing gloves, fill a window box or large tub with earth. You could add compost to the earth and mix it in. The container should be about three-quarters full of earth.

2 Scatter wild flower seeds over the soil. You can buy seeds such as daisy and bird's-foot trefoil at a nursery. Do not dig up wild plants. Cover the seeds with more earth.

3 The seedlings will come up in a few weeks. Water the young plants regularly. As the plants grow, notice which insects visit and feed on your plants.

Plant a butterfly garden

1 Grow plants from seed or buy young plants. Wearing gloves, dig over your chosen patch of earth with a trowel or a spade.

2 Break up any large clods of earth with a rake or spade. Now rake over the top of your plot so the earth is evenly spread.

3 Dig small holes for your plants with the trowel. Place the plants in the holes and press the earth down firmly with gloved hands.

MATERIALS

You will need: packets of seeds or young plants, gardening gloves, trowel, rake, watering can, notebook, pencil, field guide.

4 Water the plants well. They will need to be watered regularly through the spring and summer. The Sun will scorch wet leaves during the day, so water plants at dusk.

5 Record which butterflies you see visiting your flowers. A field guide will help you identify them. Which species prefer which flowers? Which is the most popular plant?

Butterfly bush
Buddleia is a very popular plant with butterflies and so has gained the nickname *butterfly bush*. It has masses of sweet-smelling mauve or purple flowers. Butterflies attracted to buddleia include small tortoiseshells (*shown here*), peacocks, painted ladies, commas and red admirals.

INSECTS OF THE NIGHT

M OST insects are active only by day and rest at night. But some kinds hunt and feed in darkness, or at dawn and dusk. Moths are closely related to butterflies, but unlike butterflies most are active at night. Moths are very successful insects – there are more than 100,000 kinds of moths compared to about 15,000 kinds of butterflies. Animals that are active at night are called nocturnal. Moths and other nocturnal insects have senses suited to the darkness. They cannot communicate with other insects through colour, since colours cannot be seen at night. So night-flying insects use other signals, including smells and sounds. Some moths use pheromones, or powerful scents, to communicate with the opposite sex. Other insects, including crickets and cicadas, send out sound signals to attract a mate.

Signalling with light
Glow worms communicate with light. To attract a mate, the female glow worm gives off a greenish light from her abdomen. The light is produced by a chemical reaction.

Mating song
The song of the cicada is a common sound in warm countries at dawn and dusk. These insects produce a stream of high-pitched clicking sounds using special muscles on their abdomens. Male cicadas sing to attract the females – the song of some can be heard up to 500 m away.

Fireflies use light to communicate. The males signal as they fly overhead, looking for a mate. The females see the signals and flash back. Each species of firefly has its own signal.

Cicada

Mighty moth

Not all moths are dull in colour. The spectacular atlas moth has beautiful markings on its wings. The atlas moth is one of the largest species of moths – it can have a wingspan of up to 25 cm. When a moth is at rest it usually spreads its wings out at each side of its body. Most moths have stout and hairy bodies. Some have eyespots on their wings to confuse or scare off hungry predators.

Atlas moth

Tiger moth

Confusing clicks

Tiger moths have a unique way of foiling bats hunting at night. Bats hunt by echo-location. They send out a stream of high-pitched squeaks and clicks that bounce off their prey. The bats listen to the echoes and can pinpoint the location of an insect. The tiger moth, however, can produce its own stream of clicks. These jam the bat's echo-location and confuse the hunter.

Moth antennae

Moths often have large, feathery or fern-like antennae. These are very sensitive to smell. A female moth attracts males by producing a special scent. Some male moths have such sensitive antennae they can detect the scent released by females up to 11 km away. When they are close to the female, they are able to compare the scent reaching each antenna to home in on her.

MOTH WATCH

You will need:
gardening gloves, trowel,
small torch, camera.

Moths, like all insects, are cold-blooded creatures. Most cold-blooded animals are active only during daylight hours, when the sunlight warms their bodies. So how do moths get warm enough to fly at night? Some moths have furry bodies, which hold the heat that the insect absorbs during the day. Even so, they still have to warm up their muscles before they fly. They do this by quivering their bodies and vibrating their wings before take-off. This movement generates heat.

How can you attract insects to study them when it is dark? One way of attracting moths is by using light. Moths are attracted by bright lights such as street lamps, and will fly towards them. Set up a torch in the ground and wait to see the moths arrive. If you have a camera you could try photographing the insects lit up by your torch. Another way to attract moths is by providing tasty food. Like butterflies, moths feed on liquid food, which they suck up with their proboscis, or tube-like tongue. A mushy mixture of fruit and sugar will bring them fluttering round.

Torch-light attraction

1 Wearing gloves, dig a small hole in the garden with a trowel. Do this in daylight and ask an adult's permission first before digging.

2 Check your torch fits the hole. At dusk turn on the torch and put it in the hole. Fill any gaps with earth to hold the torch in position.

3 Step back and watch the moths flutter round the light. You could try taking flash photographs of the insects with a camera.

A sweet moth feast

1 Begin by measuring out the brown sugar with a spoon into your scales or a measuring bowl. You will need about 500 g of sugar. Transfer the sugar to your mixing bowl.

2 Add the overripe fruit to the mixing bowl and mash it with a fork. Keep mashing until the fruit has become a pulp. Add some warm water until the mixture becomes runny.

M A T E R I A L S

You will need: brown sugar, spoon, weighing scales, mixing bowl, soft overripe fruit, fork, warm water, paintbrush, torch, field guide.

At night you often see moths and other insects fluttering around street lamps. They find their way in darkness by using the Moon as a guide. They flutter round street lamps because they mistake the light for the Moon and lose their sense of direction.

3 Paint the mixture on to a tree trunk or fence post. Return when it is dark. Take a torch to help you see the moths feeding, and a field guide to help you identify them.

INSECT FRIENDS AND FOES

INSECTS are vital to life on Earth. All around our planet they do many helpful jobs. They pollinate flowers and provide food for much larger creatures. They eat dead plants and animals, helping to return their nutrients to the soil. Silkworms and honey-bees are so useful to humans they have been kept for centuries. But other kinds of insects do great harm. Fleas and cockroaches infest our homes. Other insects cause serious damage to crops, bringing hardship and even starvation to some areas. Termites and some beetles eat timber, destroying living trees and wooden beams or furniture. Flies such as mosquitoes, which feed on blood, are some of the deadliest animals in the world. They pass on diseases such as malaria, which can be fatal.

This fly is feeding on ham in a sandwich. Flies also feed on animal dung and rubbish. They transfer germs to our food from their feet and mouthparts.

A nutritious food

People eat insects in many parts of the world. They are a nutritious food that can be served in many different ways. This man is selling fried grasshoppers on a stall in Uganda in Africa. In Asia, stir-fried locust is a delicacy. The diet of some aboriginal Australians includes bogong moths and juicy grubs, and in Corsica and Sardinia people enjoy a special cheese with maggots.

FACT BOX

• Cochineal insects from Central America are crushed to provide a red dye that is used to colour food. The Aztecs used this method to colour food 600 years ago.

• In Africa tsetse flies spread sleeping sickness, a deadly disease. In South America assassin bugs transmit Chagas disease.

Pest controller
Ladybirds eat aphids and other sap-sucking bugs that attack plant stems. Some farmers release ladybirds on to their crops to keep down the pests.

Crop eater
These Colorado beetles are feeding on potato plants. In North America in the 1800s, Colorado beetles caused great damage to potato crops. Now these insects are killed by powerful chemicals called insecticides, which are sprayed from the air on to the crops.

Silkworm farms
These silkworms are about to spin cocoons and pupate. In many parts of Asia, silkworms are kept on farms. The caterpillars are fattened on the leaves of the mulberry tree. Silkworms living in the wild in China and India produce wild silk. Silk is the strongest natural fibre.

Silk-maker
Some kinds of insect larvae spin a silken cocoon and pupate inside. Silk moth caterpillars, called silkworms, spin a strong, fine thread in a web around their bodies. The cocoons of some species are collected and the fibres unwound to make natural silk cloth. The holes in this cocoon show that parasites have attacked it to prey on the pupa inside.

INDEX